Preface

I want to dedicate this book to my God for giving me the wisdom & strength to make this book possible. I also want to thank my family and friends for being there with me patiently through this process. Finally, I would like to thank my mother for giving me guidance and teaching me how to stay focused even when things seem very difficult.

I thank you all!

Chapter 1

Win Lottery

This book is here to show you how to make consistent money through the lottery. This book is proof that you can win pick 3 every single day. I will begin right away to show you how to do it.

People are used to accepting everything that is in front of them. The odds of winning pick 3 lottery is 1:1000. This means that there is a total of one thousand combinations of three numbers derived from zero through nine.

The pick 4 and other lotto games are probabilities. The probable odds of winning the pick 4 and other lotto games are high . The winnings might be more on the surface. The reality is that it is more difficult and requires a whole lot more capital to play those.

The best bet you can make in any lottery game is wagering on the lowest odds available in the market. The pick 3 lottery does not offer the lowest odds, contrary to what many believe.

Most bettors play based on the potential huge payout. It would be very nice for everybody to win the million dollars but the fact remains that it is simply not going to happen. The percentage of players that do not win the lotto is higher than those that do.

The most popular games in the lotto market are the games with higher probabilities and the least popular are those with the lowest odds.

The most important thing to remember from this point on is that you stand to get the most winnings from the games with the lowest odds. This book is primarily on the two lowest odds in the market, pick 3 and pick 2 numbers.

The pick 2 method will be explained further in later chapters.

You have to cover all angles when wagering on numbers to have the possibility of winning the lottery every day. You have to cover those angles with the right methods. You don't do something because others are doing it. You do not have to lose because your neighbors and friends are not winning.

There is no single reason to play the lottery besides winning. This book will give you every tool on the planet to win the lottery every day.

Chapter 2

A critical Look

The odds of winning pick 3 lottery is 1:1000.

Most people that play the lottery do not think about OBSTRUCTION before they bet.

They do not think about DUPLICATE either.

Those two things are factors of everyday life and lottery is no exception. All I am saying here is that you must remove every clutter that could prevent you from winning the lottery consistently. If there is a duplicate to every number you are wagering on, it should be of concern. The duplicate could take the odds from 1000 to at least 2000. The obstruction could take the odds to 3000 or completely stop you from winning the lottery.

If on the other hand you understand or have the right tools to overcome those obstacles you will begin to win consistently. If you can take the odds down to less than 100, you will be ahead of the crowd. You will begin to win more than your share of the lottery money.

The key question is how do you get around those potential obstacles?

This book has series 1 and 2 to enable you overcome the obstacles.

Book 1 will give you all the explanations and how to win. Book 2 will give you more winning numbers. You cannot make the best use of series 2 without reading series 1 first.

The most important ingredient is trend. All lottery numbers follow trends.

Let us go through some simple pick 3 number trends.

The first trend would be,

100

637

207

In the above trend if winning pick 3 number 100 played followed by 637 the next winning number will be 207. You can work the group in any way you like, and the result will always come out to be the same. The lead pick 3 number in the above case is 100.

If on the other hand the lead number is 001 instead of 100, the entire trend will change.

The new result will be as follows,

001

710

863

You will notice some common numbers between the two groups. The numbers 63 from the first group shifted to the last position and the common numbers 07 from 207 played in the middle position. The positions of the common numbers shifted as well.

If on the other hand the lead winning number 100 plays as 010 the rest of the group will result in completely different sets of pick 3 numbers.

The result of the new group will appear as follows,

010

763

020

You will notice that the initial group with 100 played 637 while the group with 010 played 763. If your lottery market played 100 followed by 763 instead of 637, you might think that the trend will play 207 instead of 020.

If the 763 is playing as duplicate or obstruction, you are likely going to lose the bet unless you have for knowledge of the methods.

Let me bring down the first group that played with winning number 100 and just add 1 thereby making it 101.

Please observe the difference adding just one (1) will make to the entire group.

100

637

207

The new group with 1 added will appear as follows,

101

776

502

If I add eleven (11) to the 100 instead of 1 the result will be 111. I am sure that you know by now that the entire equation will change.

The new group with 11 added will appear as follows,

111

047

415

If I subtract 1 or add any number the entire group will change. This is where the duplicate and obstacles come in.

This equation could get more complicated. This is precisely what happens to every lotto number you are betting on.

This goes to show you that if you bet on the numbers without factoring the duplicate and obstructions, your chance of winning will be very low.

If you have the tools it doesn't matter what the trend is, the winning numbers will be right in front of you.

I can take this book to endless pages if I do calculations and explanations of every pick 3 number trend. The shortest way to give you enough data to win every pick 3 and pick 2 numbers is through the book series.

They are not voluminous for your convenience.

I will put down the above groups in one table as recap.

100	001	010	101	111
637	710	763	776	047
207	863	020	502	415

The book series 2 is extension of series 1. They are written based on calculations that will capture the winning numbers regardless of what your lottery market is playing.

The Section A in series 2 is extension of section A in series 1.

The number 5 of series 2 is extension of number 5 of series 1.

There is no place for the winning pick 3 numbers to hide.

Let me remind the readers once more that most of the explanations will be in series 1.

The next set of chapters will show you the simplified ways of winning the numbers consistently. The real money is in pick 2 numbers.

Most bettors do not know the pick 2 secret.

Come in and start making money today.

Chapter 3

Lottery Binoculars

You need to always look at the pick 3 numbers properly and not miss the opportunities of winning consistently.

The real money is through playing pick 2 lottery. A lot of people may not be familiar with the term pick 2. The pick 2 lottery is employing the method of winning with two digit numbers. If you bet 123 on the pick 3 lotto while the actual result is 124, you will end up losing the bet. If on the other hand you can win money by just catching two out of the three numbers, you will come out a winner.

You can enjoy that opportunity by playing the parlay.

The parlay wagering is the method of betting on catching two out of the three numbers from the pick 3 lottery game.

The pick 3 number 123 could be played as 12X with the X representing numbers 0 through 9. If the numbers 12 show up you will be a winner. This is the most overlooked game with the highest potential of consistent winning.

The winning pick 3 number 123 is a six way position. This means that you can rotate the numbers 123 in six positions as follows,

123

231

312

321

132

213

The parlay method is even more important in cases where you feel very good about two numbers and you are looking for the third number.

The parlay bet on 12 of the pick 3 number 123 will appear as follows,

X12

12X

X21

21X

If on the other hand you are playing parlay on the entire pick 3 number 123, the bet will appear as follows,

X12

12X

X21

21X

X31

31X

X13

13X

X23

23X

X32

32X

In the above instance you will be a winner if any of the two numbers from 123 play. You will win about $50 for every $1 bet that won.

If you are betting on double pick 3 number like 223 the cost will be half and your winnings will be the same.

The real money comes when you have the tools to capture every winning number and exponentially multiplying your bets to win the most.

That $50 payout will turn out to be $500 when you put $10 instead of $1.

It is important that you pay close attention to all the explanations in series 1 before reading series 2.

The primary tool is trend. All lottery numbers must follow trend including pick 3 and pick 2 numbers.

The trend is a group of winning numbers that exhibit similar behavior. They are related in that group and play as such.

Let us examine one trend before going further.

205

650

068

842

955

You will see that the above trend played in Maryland lottery towards the earlier part of 2011. The trend started with 842 that played as a decoy. The decoy played as 392 on 2/23/2011. You will notice that the sum of 842 and 392 is the same. The sum is 14. If you take 1 (one) from 9 of 392 and add it to 3 you will get the same number as 842. This is another reason to employ careful observation. You should not be worrying about winning the 392. It is meant to be a launching pad towards catching the rest of the group in the trend.

The 392 played on 2/23/2011. They skipped one day and followed with 068 on 2/25/2011. The next winning number skipped one day as well and played 650 on 2/27/2011. They skipped yet another day and played 205 on 3/1/2011.

You can see that the numbers in the group are exhibiting similar behavior.

That goes to show you that the trend is intact. The 392 does not by any means translate that 842 is not going to play. They may not play the entire group right after each other.

The remaining two in the group are going to play. The bettor who understands that patience is part of the game stand to win the 842 and 955.

There is no way you can miss those two if you follow the methods outlined in this book. You cannot miss those two as evidenced by Maryland lottery result of 3/11/2011 that played 824 and 955 that played six months later on 9/21/2011.

The winning numbers in the group would be profitable including the 955 that played six months later. It would have taken you from February 2011 through September 2011 to have won the entire pick 3 numbers in the group.

The series 2 on the other hand would have given you the extensions of the group thereby giving you enormous advantage and the goal of winning consistently.

Chapter 4

The Parlay Lens

Now let us bring down the last trend and look at the opportunities through a different lens (parlay).

205

650

068

842

955

I have already proven in the last chapter that the above group exhibited similar behavior. This means that the numbers in the group actually serve as mirror to other upcoming numbers. The chances of two numbers in any of the groups playing within reasonable time from each other is high. I am talking within two days.

The winning number 205 is likely going to pull down another set of two numbers from 650 or 955 within a relatively short period of time. The two numbers translate to profit for bettors that play lottery parlay.

You can actually zero in on the likely winner on keen observation of the trend.

The goal from this point is to identify two numbers if you are betting on parlay based on the above group. I will touch on parlay bet briefly with the above group and then proceed to more comprehensive ways of winning every pick 3 game in front of you.

The beauty of wagering on parlay is that it gives you winnings where otherwise it might not exist. You should always endeavor to play the parlay along with the pick 3. You stand a very good chance of winning the parlay than the pick 3.

An instance of this will be from Maryland lottery pick 3 results that played 502 on 9/20/2011 followed by 955 and 084 on 9/21/2011. The 502 and 955 appeared in the above sample group. The next winning number in our group is 842 while Maryland played 084.

You would have lost on the 084 as pick 3 but win the play as parlay. You would have won the parlay as X84 and that would pay you about $50 for every $1 bet. The 502 and 955 would have shown you the direction of the trend which would have resulted in winning the parlay.

You will always win the parlay if two of the three numbers show up together. The $50 win on $1 is overlooked by many. That money will begin to add up if you get the parlay consistently and bet more on it. The $50 win would turn to $500 if you bet $10 on it.

If you have enough to capture every parlay in front of you, the real money journey begins. There is no limit to how much you can make.

The real journey begins from this point on.

Please pay utmost attention from here on!

Chapter 5

Professional Betting 1

This is where you identify what works and you stick to it. The winning lottery numbers could take different dimensions for all the reasons I have given so far. The series will capture every winning number if you observe and follow the trend.

I will show you an instance with different lottery markets to prepare you for every pick 3 and parlay game.

Let us take a look at the entire table and the potential opportunities it presents on a daily basis.

Figure 1

	A	B	C	D	E	F
1	584	884	884	984	458	505
2	968	529	529	529	959	802
3	932	459	959	459	758	331
4	035	345	445	845	037	802
5	944	163	263	363	401	540

Maryland lottery started playing the Figure 1A group above on 1/1/2011 with 584. The next winning number, 968 followed one day later and played as 698 on 1/2/2011. The third pick 3 number in the group was 932 that played as 239 on 1/3/2011.

The winning pick 3 numbers in the trend played right after one another. There are instances the group may spread over several days. The next winning numbers that played were 529 on 1/3/2011 and 459 on 1/4/2011. The Figure 1A trend calculations have pick 3 numbers 035 and 944.

If you reverse the 035 it will become 530. If you subtract 1 (one) from 530 the answer will be 529. This means that they took one from 530 and added it to 944 thereby making it 459.

A lot of bettors will miss the last two pick 3 winners. You are likely going to win at least parlay on the 459. The parlay gives you the opportunity to win if two of the three numbers shows up.

The parlay would have given you the cost of the bet plus more money in your pocket. You would have made $50 on every $1 bet winner. This is the opportunity that is not known and overlooked by many.

The winning pick 3 number 459 from Figure 1A would have been possible only if you had bet on the parlay. The 944 in column 5 would have made some money for you. This of course is based on the position that Figure 1A is the only available option.

The winning pick 3 number 529 on the other hand would have taken you to the other columns. You would have won with Figure 1B and 1D, as well as 1C if you bet on the parlay.

It is important to understand why parlay betting is the most powerful tool in lotto game. Every winning number begins with two digits. The number one is actually 01. Two pick 3 numbers for instance 123 and 456 could be three two digit numbers that look as follows 12, 34 and 56. Those numbers in turn could be part of the required six lotto numbers.

Let us go further in proving why parlay betting is essential. The Figure 1B trend started with 884. Maryland lottery on the other hand played 584 on 1/1/2011. The 584 and 884 are actually playing the same trend. The difference in the lead numbers between 884 and 584 is 3. You may not know initially that the two groups are playing the same trend until the next winning number 529 played on 1/3/2011. You will notice that three other pick 3 numbers (959, 698 and 239) played in Maryland lottery (between 1/2/2011 and 1/3/2011) from 584 prior to 529.

Those three winning numbers justify the difference of 3 in the lead position between 884 and 584. That is why the trend in Figure 1A and Figure 1B are the same. The bettor that bet parlay on row 1 would win Figure 1A through 1E as against betting on just the pick 3 that might have won only 1A.

The same advantage applies to all the columns. You will make money on 584 and 884 when you bet on the pick 3 and parlay.

If you are betting based on Figure 1E, you will notice that 458 played on 1/1/2011 followed by 959 on 1/2/2011 . The third winning number in the group 758 played straight three days later on 1/5/2011. The 758 would pay good money to all the patient bettors. That payout would be $500 for every $1 bet.

The next winning numbers in the trend will be 037 and 401. The way to bet parlay on those will be to play it for the next two days. In most cases a minimum of two digits will show up, thereby making you a winner. In the event that it did not happen (as in this case) you will put the two pick 3 numbers 037 and 401 in your watch list.

This means that you should consider betting on those any time one of the numbers play. The entire trend could begin an entirely new rotation at that point.

Another important point in this matter is that the two numbers are going to play to complete the trend. You can make the decision to continue playing the two until they drop.

To players who are new in this game, the words drop, hit , win and play could be used interchangeably.

Watch List

Consider betting pick 3 and parlay on the winning numbers below whenever one of them hits. Please bet on the next pick 3 for two days. Watch out to see if the entire group is playing and go for the opportunity

037

401

The watch list works more for the players who prefer to play only pick 3. I must state once more that you can make money every day if you master how to bet on parlay.

The two books, series 1 and 2 will cover every pick 3 lottery number in front of you.

If you bet on parlay, for instance, you don't have to wait for the two winning numbers above on the watch list. You can, however, use the two and win money consistently. In the parlay method, you will be looking out for two digits among the 037 and 401.

Let us look at some that could have made some money for you in Maryland lottery.

Maryland lottery pick 3 played the following numbers 039 on 4/8/2011 followed by 601 on 4/10/2011. You would have won money with 601. You can see the similarities between 039 and 037 and that of 401 and 601.

The two pick 3 numbers in the watch list became the guide to win on that bet.

Other numbers you would have won include 540 on 4/19/2011 and 130 on 4/20/2011, 405 on 7/28/2011 and 732 on 7/29/2011, 737 on 8/31/2011 and 406 on 9/1/2011.

The 401 finally played on 11/3/2011 followed by 038 on 11/4/2011. The trend calculation in this book gave 037 while they played 038. The bettor who played the parlay would have won on the 038.

The 401 came back on 12/15/2011 followed by 083 on 12/29/2011.

The 038 does not, however, mean that our manual trend calculation is not going to play in Maryland or any other state for that matter. You can see that evidence in DC lottery results of 2010 and as recent as DC 3 playing 037 on 1/20/2012 followed by 959 on 1/24/2012 and 401 on 1/27/2012.

You can see that DC lottery is currently playing the Figure 1E group. This same group and the other ones could be playing in your home state. The book series are made to put you in line with the trend that your lottery market is playing at any point in time.

Let us bring down the Figure 1 table and go through 1F column because the trend is different from the other columns.

Figure 1

	A	B	C	D	E	F
1	584	884	884	984	458	505
2	968	529	529	529	959	802
3	932	459	959	459	758	331
4	035	345	445	845	037	802
5	944	163	263	363	401	540

The Figure 1F trend started in earnest on 1/11/2011 with winning pick 3 number 505. The next winner might have occurred if you bet parlay on the 543 that played on 1/12/2011. The Figure 1F group did not play right after each other. This group is clearly following a different trend.

You would, at this point, put the rest of the group on watch list. The parlay player on the other hand would win more money a few days later on 620 that played on 1/16/2011 following 433 that hit on 1/15/2011.

For the readers that might be tempted to ask why, the reason is because the 433 is close to the 331 in our trend. The 620 would have won based on the 802 in our unique calculations.

Once again you can see that the parlay method awards you with enormous opportunities to have consistent winnings.

The 331 is the next winning number among the group that played on 2/25/2011. You might have bet parlay based on 802 through 2/27/2011. If the numbers do not play within two days then you should leave it and pursue the next opportunities.

The reality is that you are not going to worry about the next number to bet on if you have the two books in the series. The books are meant to zero in on any pick 3 number your lotto market is playing.

The Figure 1F has 802 as the repeat candidate. If you go back towards the end of 2010, you will notice that the 802 played several times. The same number came back and played as 280 on 5/17/2011. It repeated three more times on 6/5/2011, 6/18/2011 and 6/28/2011.

This is one instance that bettors that play pick 3 stand to make a lot of money.

This means that the other pick 3 winning numbers in the group are going to repeat several times. You do not need anything else to remind you that you have gold in your hand.

The trends are groups of winning numbers that exhibit similar behavior. You should put the rest of the group on your watch list with a special mark to remind you. I would not have any problem playing the entire group from this point on.

One other thing worth noting is that this trend works in every lottery market. You should jump on them anytime your state starts playing the numbers. If you are in doubt check the numbers in the groups with past lottery results.

You can check three different states and go back one hundred years. The result will be consistent.

The mastery of parlay method gives you the opportunity to win every single day.

Figure 2

	A	B	C	D	E	F
1	756	657	652	381	418	332
2	934	228	651	652	431	554
3	941	520	156	392	743	179
4	257	532	543	948	888	015
5	333	739	746	457	953	718

New Jersey lottery pick 3 started playing the Figure 2A group on 2/21/2011 with the 756 playing as 576. The parlay bettor would have made some money with 336 that played on 2/23/2011. The wager would have netted some money based on the 333.

It is important to note that the trend could follow any direction. The next winning pick 3 number in the group that played is 941 on 2/26/2011. The 941 played two days earlier as 194. New Jersey lottery played 756, skipped 934 and played 941.

The parlay bettor would have won again since two common numbers, 94, are present between 941 and 934.

The next pick 3 number in the group that won was 333 on 2/28/2011. Once again the parlay bettor would win on 520 that played on the evening result. The player would bet based on the 257 that is next to 333 based on the trend.

You can see why the parlay player could easily win the lottery every single day.

The trend played 257 few days later on 3/3/2011.

If you take a critical look on the trend from the beginning of March 2011 you would have seen the 257 coming. You will notice the common numbers among the pick 3 numbers that played from 3/1/2011.

New Jersey lottery played 433 on 3/1/2011, 134 on 3/2/2011, 014 on 3/3/2011 followed by 257. You can see the common two digits among the above winning numbers and the ones in Figure 2A.

The pick 3 winning number that is left to complete the trend would be 934. The parlay bettor on the other hand would have continued the winning streak with 893 that played on 3/5/2011. The beauty of the parlay method is that you do not have to be concerned about when the number is going to play.

You should always play the number as well as the parlay. The 893 would have paid off based on the 934. You can see the common two digits 9 and 3 between the two. The 93 appeared within two days from the 257. The two days is the common rule. If the parlay does not play within two days please move on to the next opportunities.

You will not bother about that once you have book 1 and 2.

The 934 played on 3/10/2011 and repeated on 3/11/2011. It is important to remember that 941 played two days earlier as 194. You can see that the 934 is actually pulled down by 941. The New Jersey lottery players would have made good money on Figure 2A within a relatively short period of time.

Now that the rotation is complete you should put down the following statements in your watch list.

Watch list

New Jersey pick 3 numbers 941 and 934 move together. Remember to look out for those two and consider betting on them any time the trend comes back.

The Figure 2B column is playing the same trend with that of 2A. They started within one day and ended the same way. Figure 2B played 228 on 2/20/2011 followed by 657 on 2/21/2011. The winning pick 3 number separating the group is 520 that played by itself on 2/28/2011.

The trend completed with 532 and 739 that played on 3/10/2011. It is important to note that the parlay bettor would have won more money. The parlay rule of playing within two days of prior result worked in every case here.

The parlay bettor would have won more with 254 and 253.

New York lottery numbers definitely played Figure 2C trend starting with 652 that played on 2/1/2011. The winning number in the trend is 651 that played on 2/3/2011. They repeated the 651 as 156 on 2/11/2011.

They skipped 534 and played 647 on 2/12/2011. You must know that at this point the 543 is going to play to complete the rotation. The 345 played followed by 165 the same day on 3/22/2011.

The parlay method would have paid a lot more because it did not violate the two day rule. Please feel free to check it. Please endeavor to check the numbers you are playing against other rows and columns as well as other state lottery results.

Once you master how to read the table you should be able to accurately call winning lottery numbers of several states at the same time.

New York lottery number fans would have made serious money from Figure 2D in one month. The trend started by playing 381 and 652 on 2/1/2011. The next winning number among the group that played was 948 on 2/28/2011.

Please note that the parlay betting fans would have had two additional winnings with 946 on 2/2/2011 and 045 on 2/4/2011. The good thing about the parlay is that you do not have to wait for 457 in this case to play. It is going to play to complete the cycle.

The 457 should go into the watch list pending the time the trend comes back.

The Figure 2C and 2D clearly shows that you can be following several groups simultaneously. In this case it would have benefited you enormously. New York lottery played the two groups in a very short period of time.

Take another look at the two groups and go through New York number results of February and March of 2011. The winning numbers actually played within two weeks. That is a lot of money for every good observer. The parlay bettors would have made much more.

The best part of Figure 2D is that the entire group actually played straight. That is very rare over such a short period of time considering the fact that each one of those pays upwards of $500 for every $1 won.

New York lottery started playing Figure 2E trend on 1/29/2011 with the winning numbers 418. The parlay bettor would have won some money the next day with 953 based on the New York result of 739. You can see the common numbers 39 among both. This is possible despite the fact that 953 is the last winning number in the group that played, as of the time of writing this book.

The next winning number in the group that played was 743 on 2/8/2011. They came back and played 431 on 2/11/2011. The Figure 2E above shows you the actual trend. The next winning number that played was 953 on 2/15/2011.

The only winning number that is left to complete the cycle is 888. You may consider playing the 888 and parlay as well without waiting for any special date. The reason for that decision would be that the ones that played already did within a very short period of time.

The second reason is because 888 is triple number. Your chance of getting 88 in a front and back pair is high. In the event that it comes out as 888 you will not just win the pick 3 but you would win the parlay as well. You would play the parlay in this order: 88X and X88.

They played 088 ten days later on 2/25/2011. That would be a profitable bet.

New York numbers played Figure 2F in less than one week. The group is laid out in the order of the trend. New York lottery played 332 on 2/12/2011 followed by 554 on 2/13/2011. They skipped one day and continued by playing 015 on 2/15/2011 and 179 on 2/16/2011.

It is important to note that they can jump one number and come back again later to complete the rotation. The last number in the group is 718 that played as 781 on 2/18/2011.

The above trend played in New York in a very short period of time. They may not necessarily play in your local lottery market the same way. Your state might decide to spread them over a period of one month or so. The larger point here is that you have the group that moves as a trend right in front of you.

Watch out for the trend and make money with them in your market. The knowledge of the trend gives you advantage of not worrying about the odds. The parlay bet gives you the opportunity to make a whole lot more money.

Chapter 6

Professional Betting 2

Please continue reading after studying Figure 3 through Figure 6 to get a clearer picture.

Figure 3

	A	B	C	D	E	F
1	847	229	788	370	786	806
2	847	334	699	960	790	652
3	565	022	973	476	032	749
4	171	478	687	678	277	267
5	137	841	493	877	762	833

Figure 4

	A	B	C	D	E	F
1	110	806	483	527	392	962
2	106	021	805	569	933	732
3	710	896	430	848	087	130
4	680	559	870	277	239	222
5	764	145	719	220	176	655

Figure 5

	A	B	C	D	E	F
1	955	412	211	941	559	722
2	333	214	844	888	081	686
3	572	577	341	112	758	456
4	863	784	485	284	853	146
5	142	044	821	716	141	453

Figure 6

	A	B	C	D	E	F
1	674	392	800	817	278	066
2	989	774	797	946	383	225
3	823	868	731	207	787	709
4	821	885	766	371	761	267
5	713	440	782	628	421	473

The lottery winning numbers come in different shapes and sizes. The most important thing is knowing that all the winning numbers must go through a given trend. The Figures in this book as well as series 2 are meant to cover every trend is playing.

A look at Figure 3A through Texas lottery pick 3 will show you that the trend played in Texas beginning with pick 3 number 847 on 3/9/2011 and repeated on 3/12/2011. It is important to note that 478 that played on 3/2/2011 is not part of this trend.

They played the same way the trend calculation in this book showed the winning pick 3 numbers.

The 171 followed on the one month anniversary of 847 by dropping on 5/12/2011. This is not by accident. It is only possible through lottery manual trend calculations.

This group played over extended period of time. The bettors that wager on parlay would follow that method as I have been explaining in this book i.e. bet parlay for two days and move on to the next opportunity.

The other tables in this book will give you ample winning numbers based on the trend.

You will put the other winning pick 3 numbers, 565, 171 and 137 in your watch list.

The next winning pick 3 number in the group is 565 that played on 12/21/2011 and repeated on 12/22/2011. You can see that the group is exhibiting the same behavior. The 847 played twice followed by 565 that did the same.

The 565 must have enabled good observers to make good holiday shopping.

The next winning number in the group that played is 137 on 12/30/2011.

The parlay bettors would have won more as usual based on 172 that played following 565 on 12/21/2011 as well as 732 that played after 137.

The Figure 3A group is playing over extended period of time but the entire group is poised to pay good money since they are repeat candidates.

The power of manual trend calculation is the ability to capture the winning numbers regardless of how they play them.

Please take a look at Figure 3B to buttress my argument on patterns the trend could play. Texas lottery pick 3 played the Figure 3B trend starting with 292 that played to end January 2011 on 1/31/2011. The trend continued with 220 that played on 2/1/2011. The trend paused for one month and came back with 334. The 334 is actually pulled down by 229.

You can see the actual trend calculation on Figure 3B.

The group resumed on the one month anniversary of 220 with 334 playing on 3/1/2011. You can see the behavior between the groups. The dates are even important in a lot of cases. The next winning pick 3 number that played was 478 on 3/2/2011. This will be just one day apart.

The players that bet on parlay would win more in this case.

The next winning pick 3 number that followed was 229 on 3/10/2011. This will be a repeat. The 229 once more pulled down 334 on 3/15/2011. You would think that the trend is done at this point.

The next winning pick 3 number that played was 220 on the three month anniversary of 334. The 220 played on 6/15/2011. You can see the power of manual trend calculation. A good observer must have taken note of the 15th of those two months.

The pick 3 number left to complete the cycle will be 841 of Figure 3B. That number took a few months before playing as 148 on 11/15/2011. That will be yet another 15th of the month. You can see the precision.

The more important point is knowing that the Figure 3B trend exhibited similar behavior. If you watch out for the group in your home state you could make some big money. The trend in this case virtually eliminates the need for you to worry about the odds.

The key to consistent winning is keen observation. Let us take a quick look through Figure 4A with the Ohio lottery pick 3.

The Figure 4A trend started playing in the Ohio lottery pick 3 with 110 that played as 011 on 2/3/2011. The next day they played 471. The 471 would have paid some money to the parlay players based on 764. This is meant to give you back the cost of wagering plus some profit. You can see the increase in the potential winnings based on the amount that is placed on the bet. You will make that decision once you master the method.

The trend resumed towards the end of the month with winning pick 3 number 106 that played as 061 on 2/28/2011. The next day Ohio lottery played 740 on 3/1/2011 followed by 680 that played straight on 3/2/2011.

You can see that the groups played right after each other. The only pick three left to complete that rotation is 764 which they played on 6/3/2011 as 467.

Please always try to use other trends or groups to help you capture the winning numbers in any market you are playing.

An instance of such scenario will be the trend in Figure 5A and 5B that played in New Mexico.

The two groups played within reasonable time from each other.

The Figure 5A trend started playing in New Mexico lottery as 142 that played on 1/8/2011. In Figure 5B the same trend played. In the case of 5B the manual trend calculation showed the winning number as 412. The pick 3 number, however, played in the format Figure 5A had it.

The two parties would have made money regardless.

The Figure 5A trend continued by playing 386 on 1/21/2011 followed by 955 on 1/25/2011. New Mexico lottery played 672 on 1/26/2011 as against the 572 that our calculation is showing. This is one more reason it is important to consider playing parlay. The parlay player would cash $50 for every $1 bet.

The 572 is going to play at some point. You are looking at maximizing your winnings. You do not necessarily have to wait for 572.

In any event, you can tell that the group is playing within a reasonably short period of time. The only winning pick 3 number left to complete the cycle is 333. That winner played one day later on 1/27/2011.

The figure 5B on the other hand had the 412 that repeated. That trend calculation clearly showed both winning numbers. The first 412 played on 1/8/2011 and the accompanying 412 repeated on 2/16/2011.

The next winning New Mexico pick 3 number was 577 that played on 2/22/2011 followed by 784 two days later on 2/24/2011.

The pick 3 number left to complete the rotation is 044. The 044 waited few months and played on 8/10/2011.

In the case of 5A they played 672 against our 572 and continued the trend with 333. It is important to note that 572 pulled down the 333. In the case of 5B they played 577 straight and that winning pick 3 number pulled down 784.

If you are playing the Figure 5A trend through 572, you will be able to switch to 5B based on the 577 and end up winning the 784 based on the 5B trend.

The trend could change in some cases, thereby producing a completely different set of numbers. This book series is meant to capture all pick 3 and pick 2 lottery trends.

In the above case if the 577 is in series 2, the only way you will be able to win the 784 is by looking through series 2.

There are instances when the entire group will play within a short period of time. Those instances should be taken advantage of and celebrated. One of those is that of Figure 6A that played in Florida lottery Cash 3.

Florida lottery started playing the trend on 4/22/2011 with the winning Cash 3 number 821. You must recognize that each group is actually a circle. The trend could start from any of the winning numbers and continue in either direction until the rest of the cycle completes.

That rotation will start again at some point and go through the same process. The position of having the entire group worked out saves you the trouble of being confined within the odds limits. One important thing to bring to the table is patience. It works well with wagering.

The Figure 6A started with 821 on 4/22/2011 followed by 713 that played as 173 two days later on 4/24/2011. The trend continued the next day by playing 989 on 4/25/2011 followed by 674 on 4/26/2011. You can see that the entire group is playing almost right after each other. They completed the cycle by playing 823 on 4/30/2011 to cap the month.

The Florida lottery good observers would have made good money that month. The same group could play in your local lottery market. Please do not miss them when that opportunity arrives.

Georgia lottery Cash 3 played group 6B trend starting with 392 that hit on 2/26/2011 followed by 774 on 2/28/2011. The trend continued by playing 868 on 3/1/2011 followed by 885 that played as 588 on 3/7/2011. They completed the trend by playing 440 as 404 on 3/15/2011.

This book gives you the actual trend calculation as they are and give you the numbers the way they played. They are going to play in some other states the way the actual calculation is. The parlay method gives you a chance to win more regardless of how the numbers play.

Georgia Cash 3 played Figure 6C similar to that of 6A that Florida Cash 3 played. The winning Cash 3 numbers played in a very short time.

You must pay close attention to recognize the trends when they are playing. The winning numbers will always be there for you to notice early enough and make good money on close observations.

Georgia Cash 3 started playing Figure 6C trend with 731 that played as 173 on 12/26/2010.

That winning Cash 3 number played on the boxing day. You must be prepared for the match. The trend continued right from 1/1/2011. That of course is new year. You can see that this trend started by making use of two special dates.

The next winning Cash 3 number among the group that played was 797 on 1/5/2011 followed by 782 on 1/14/2011. The manual trend calculation clearly shows you that they skipped 766. At this point you know they will come back and play the 766 to complete the rotation.

A lot of players might have bet on 766 after the 797 played. The would have because that is actually the order the real trend is. If that was the case the bettors who played parlay would have won with the 678 that played right after 797. They would have made money despite the fact that 766 did not play.

The 766 played as 667 on 1/22/2011 to complete the cycle.

Another important key to winning consistently is keen observation.

The readers that got this far in the book please pay close attention to Figure 9 and figure 10 tables. The methods there will be quite different but powerful.

The ultimate goal of this book series is for you to win lottery every day.

Figure 7

	A	B	C	D	E	F
1	987	091	477	770	088	425
2	798	220	369	844	880	477
3	269	104	784	718	014	378
4	485	341	432	270	750	425
5	375	155	218	628	602	286

Figure 8

	A	B	C	D	E	F
1	114	319	271	250	325	255
2	325	667	118	260	621	099
3	217	767	468	591	621	729
4	980	341	914	890	734	185
5	780	262	416	687	762	227

As I have stated several times in this book the trend are group of winning numbers that exhibit similar behavior. They move in tandem when the trends are playing. The winning pick 3 numbers in any particular group or groups are actually like a circle.

The winning numbers will continue to play in that rotation. The difficulty with most lottery players is doing the manual calculations and capturing the winning numbers in each circle. You have the numbers in this book. You will have the complete trends to win every lotto in all two series of this book.

A quick look at Figure 7A through Virginia lottery pick 3 will confirm once more that the rotation could start from any point but will ultimately complete the cycle.

The Virginia lottery pick 3 started playing Figure 7A with 798 that played on 1/2/2011 followed by 987 on 1/9/2011. The manual calculations will show you the way the actual trend calculation is. In some cases your local lottery might play the numbers per the trend calculations.

The next winning Virginia pick 3 number that followed one day later was 485 on 1/10/2011. They came back a few days later and played 269 on 1/15/2011.

The trend completed by playing 375 as 735 on 1/18/2011. If you take another close look at the winning numbers, you will see that the players who bet on parlay would have won many more. All you need to do is go through the winning numbers and check what played within two days from that result. The ones with two common numbers would have won some money.

Virginia pick 3 lottery played Figure 7B trend starting with 091 on 1/4/2011 followed by 220 on 1/7/2011. One week later they played 155 and 104 the same day of 1/14/2011. The 155 played as 551 and 104 played as 041.

The only remaining winning pick 3 number to complete the trend would be 341.

Let me show you how I would bet on this play in Virginia if I was playing it and the reasoning behind it.

The only winning number left at this point is clearly 341. I would put down 341 as one of the games I am going to bet on. I would again put down 041 since the two have two common numbers (41) among them. I am putting down 041 in case it plays as repeat against the 341 that my calculation is showing.

My next option will be betting on the 341 as parlay. I would have won the parlay based on 436 that played on 1/15/2011. You can see that I would have made money despite the fact that 341 did not play and 041 didn't repeat.

I was looking at capturing minimum of two common numbers to win the parlay. In this case the 34 made it in the 436.

The 341 did not play until 5/23/2011. The bettors who play parlay would continue to win.

The parlay gives you the opportunity to win money where otherwise it would not be possible.

Let us take a look at the Figure 7B trend through a different trend. The reason for that is to actually show you that once the calculation is correct the trends works in every lottery market.

Texas lottery pick 3 played the same trend starting with 155 on 3/17/2011 and repeated on 3/29/2011. You may remember earlier that I stated that the trends could start from any of the winning numbers in the group. The reason for that is that the trend is actually like a circle. The winning numbers will always rotate within that circle anytime the trend is playing. That is precisely what makes the winning pick 3 numbers good in every lotto market.

The next winning Texas pick 3 number in the trend that played was 104 that played as 401 on 5/6/2011. The trend continued by playing 341 on 5/10/2011 and repeating the same winning number on 5/23/2011. The next winning number in the group that played was 091 on 6/2/2011 followed by 220 on 6/15/2011. The pick 3 numbers in their respective groups exhibited similar behavior. They continued by repeating 091 on 7/9/2011.

You can see that the same winning pick 3 numbers from Figure 7B that played in Virginia did the same in Texas. I implore you to check those in your state's past lottery results. If it has been a while it played in your market, make note of them along with the other ones and look forward to winning good money.

Please apply the same methods in your local lottery market. Make use of the rest of figures 7 and 8 and then go to the Super Methods.

Those you will find in Figures 9 and 10.

Chapter 7

Super Methods

Figure 9

	A	B	C	D	E	F
1	750	750	750	750	750	750
2	970	443	077	177	020	834
3	260	069	277	817	418	443
4	601	118	113	383	364	090
5	246	278	473	973	078	689

The Figure 9A trend played in several states including Maryland pick 3. Maryland lottery started that trend with 750 that played on 3/21/2011. The next winning pick 3 number in the trend that played was 246 that played as 462 on 3/24/2011.

You may naturally think that the next winning pick 3 number based on that direction would be 601. The next winning Maryland lottery pick 3 number was actually 970 on 3/27/2011. You must remember that the winning numbers in the group are like circles.

You local lottery could decide whichever direction the games would play. The next winning number that played after 970 is 260 on 3/28/2011. That would be just one day apart.

This group played 80% of the winning numbers in the group in just one week. Those winning numbers played from 3/21/2011 through 3/28/2011. The last winning pick 3 number among the group is 601 that played on 4/10/2011. You will notice on close observation that 970 actually repeated. It did on 4/9/2011 thereby paving the way for 601.

The Figure 9B is another case in point on the overall argument. That group started playing in Maryland with the 069 that is in the third row. Maryland lottery played 069 as 906 on 3/15/2011 followed by 443 that played as 434 on 3/17/2011. The lead pick 3 number in the trend, 750 played on 3/21/2011.

It is very critical that you apply good observations all the time. There are bettors who will be looking at only 750 and end up missing the rest of the group. This could happen despite the fact that the entire group played within the same short period of time.

The next Maryland lottery winning pick 3 number among the group was 118 that played as 181 on 3/25/2011 followed by 278 one day later, on 3/26/2011.

The Figure 9A and Figure 9B had the same lead pick 3 number 750, yet they followed two different patterns. The Figure 9B followed right after each other from 069 while the Figure 9A did not.

The two groups played around the same period of time. The last winning number from each group gives you idea of the trend that is playing. The winning pick 3 number 118 for instance means that your bias should be on Figure 9B. That bias would lead you to play the 278 especially in this case, since 278 is the only winning pick 3 number among the group that is yet to play.

In absence of Figure 9A you might have a more difficult time working out the last winning pick 3 number 278.

This is another reason to collect all the series. It is not possible to have all the winning numbers that started with 750 in this group. The equation for the rest of the group will change every time one or more numbers is added to any group.

Take a look at Figure 9C and 9D. They are the same group of number before I started doing the calculations. The two groups changed course from the 077 and 177 of Figure 9C and 9D respectively. The trend works everywhere once the calculation is correct.

If you make one mistake for instance writing 111 instead of 112 the entire calculation is going to be wrong. The need to avoid such mistakes is important especially in cases where you have limited choice of trends.

You do not have to worry about that since the trends are already calculated in this book series.

The other beauty of working with the right trends is that they work everywhere. If you are in doubt take any of the groups, go back and check them over one hundred years or however long you can. They are going to be consistent. Please check them with different state lottery results.

The trends are groups of numbers that exhibit similar behavior. You are confined within the odds of the winning numbers you are working with in the case of this book that would be five winning numbers in each group.

You have odds calculation of one thousand when you don't have anywhere to start and end up chasing every number. That is a recipe for compulsion.

You can take any particular group in this book and see how long the entire group played in any rotation. You will find out as I have already shown that the groups played over a relatively short period of time. The way to really know if you could come out ahead is by checking prior history of that trend through some state lottery prior results.

The pick 3 number won straight on $1 bet pays you about $500. It will take 500 games to exhaust $500 at the rate of $1 per game. The trend do play over short period of time and in most cases within a period of one month.

If you take one particular group and start playing once the trend is established, you stand a very good chance of making serious money. The parlay method could catapult your winnings several times over.

Take the winning pick 3 number groups in Figure 9C and 9D and check them through Georgia lottery Cash 3 results of the year 2011. Study how the two groups played against each other and apply it in your home state or country.

Maryland lottery played Figure 9E with the same lead pick 3 number 750 but the entire group followed different trend. It is important that you take another look at the entire groups of Figure 9 with Maryland lottery result.

The groups are different from each other despite the fact that the lead pick 3 number among them is 750. Another thing that is equally important is that the winning numbers played around the same time period. This enables you to enjoy consistent winnings.

The trend could change at any point and you will still be able to capture the winnings. The bettors who play parlay will definitely win a whole lot more.

The Figure 9E followed different trend yet the group played within the same time frame.

The Figure 9E group started playing with 750 on 3/21/2011 followed by 078 that played as 870 one day later on 3/22/2011. The next winning number based on the trend is 364 that played on 3/27/2011. The trend continued by playing 020 on 4/17/2011 followed by 148 on 4/20/2011 to complete the trend.

The group could start playing in your home state at any point. Please take note of the trends and make some money with them.

There is no other lottery book out there that has the concentration of trends with the same lead number. The difficulty is understanding how to do the calculation. The groups in Figure 9C and 9D is just one tiny example. You can see the difference beginning with the winning numbers in the third row (277 and 817). The rest of the groups changed. This is just as a result of 077 in Figure 9C that played as 177 in Figure 9D.

After studying the groups with Maryland lottery apply the same methodology to your own state lottery pick 3 games. The numbers work in every market because they are manually calculated based on trend.

Figure 10

	A	B	C	D	E	F
1	751	752	753	666	777	888
2	830	712	957	856	663	962
3	218	097	590	800	453	055
4	123	235	888	425	723	547
5	681	058	108	732	318	328

The Figure 10A through 10C started the trends with 751, 752 and 753 respectively. The Figure 10D through 10F on the other hand started with triple winning numbers 666, 777 and 888. The rest of the groups are definitely going to be different from each other.

There are times your local lottery might play 751 followed by 830 as in the case of Figure 10A. In that situation you would naturally expect them to play 218 or 681. They are not necessarily going to do that all of the time. It could be a situation where the 218 or 681 might take another week or two to play.

That doesn't mean that the trend is violated or not intact.

Let us assume for instance that they chose to play 097 from Figure 10B or 453 from Figure 10E. In the last two instances it gives you the opportunity to follow what they are playing by focusing your attention on 10B or 10E as the case may be.

There are instances when they will play pick 3 numbers that are completely different from the trend that you are following. An example of that would be if they play 848 or 392 for instance. You would not find those two pick 3 numbers in Figure 10.

The best approach in this instance would be to look through the rest of other groups in this book series. You will find 848 in Figure 4D and 392 in Figure 2B. The primary reason those two could come in play is because they are in the third row being the same position as that of 218 from Figure 10.

You must realize that the winning numbers in the entire book series are calculated to move in tandem. You can use any group to recapture the trend once the winning trend shift. The likelihood of that trend shifting into another group in the book is high thereby giving you the opportunity to make money consistently.

The bettors who enjoy playing the parlay will actually be looking to identify two common numbers from the groups that will correspond with the trend they are playing.

The mastery of parlay wagering is the key to winning every day.

One instance of the above scenario would be the trend that connected Figure 10 and Figure 1 with Tennessee pick 3 lottery prior results.

The trend started with Tennessee lottery playing 751 and 752 from Figure 10A and 10B on 6/8/2011. The next day the trend shifted to Figure 1E by playing 458 in the same first row position on 6/9/2011. This clearly would be one of those situations you need to keep watchful eye on both groups to see the direction the trend would establish.

In this scenario you should not jump in right away because the 751, 752 and 458 are in the same first row despite the fact that they played one day apart. If the 458 had been on the second row you would then put your focus more towards that direction.

Your goal should be to wager only to win. This is serious business contrary to what many would think or believe.

They came back two days later on 6/11/2011 and played 599. You can now see that the trend is squarely on Figure 1E. Tennessee pick 3 lottery continued the trend by repeated 752 on 6/20/2011. The next winning pick 3 number based on the trend in Figure 1E is 758 that they played one day later on 6/21/2011.

The next two winning pick 3 number among the Figure 1E group to complete the cycle are 037 and 401. The 758 played on 6/21/2011 and the next potential based on the trend is 037 followed by 401.

Tennessee lottery, however, played 004 on 6/22/2011 followed by 637 on 6/23/2011. The bettor who played parlay would have won those two despite the fact that 037 and 401 are yet to play. This is another reason why it is important to consider putting the parlay among your best plan.

There are times one or more numbers could change. You stand a good chance of winning the parlay when one digit is altered and move to another related trend when more than one number is different.

The trends in this book are manually and painstakingly worked out to give you certain advantages.

They tend to play over a relatively short period of time.

This must be celebrated because of the fact that one dollar bet won on pick 3 pays you about $500. If the winning numbers in a given trend plays before the length of time it would take to exhaust the potential winnings, you stand to make good money.

The trend follows a certain pattern.

You do not need to worry about the odds beyond the given trend you are playing. It is easier to play a group of five pick 3 numbers in any established trend than chasing combinations of 1000 pick 3 numbers.

The trend is good in every lottery market including yours.

You can go and back test the winning pick 3 numbers in any given group as far as you can.

You could win good money over relatively short period of time as established in the trends through the groups in this book including the groups in Figure 10.

The Figure 10 is definitely no exception and that position could easily be established by looking at some of the groups.

New York lottery played the trend in Figure 10A with the 751 being in the lead position. That trend started by playing 123 on 3/18/2011 followed by 681 that played as 168 on 4/4/2011. They continued by playing 830 on 5/12/2011 followed by 218 on 6/5/2011. They completed the cycle few days later by playing 751 on 6/11/2011.

You can see in the above pattern that they played the winning numbers every month beginning from March through June. You stand to win one in this case plus more with the parlay method.

New Jersey lottery pick 3 followed by playing Figure 10B over a very short period of time. The lead pick 3 number is 752 even though the trend started by playing

097 on 9/2/2011 followed by 235 on 9/7/2011 and 752 on 9/8/2011. They continued by playing 712 as 127 on 11/9/2011 and capped the trend with 058 on 10/15/2011.

You can see that the group played in just a little over one month. That would be huge money in such a short period of time. The manual calculation is capable of displaying the winning numbers in the order they are meant to appear. This means that your own local lottery market could indeed play them in that order.

The order the winning pick 3 numbers play, however, should not be your primary concern. The key is to win big with the numbers. They are already worked out and ready for you to make use of.

The rest of the groups in Figure 10 exhibited similar pattern.

You could reward yourself handsomely and consistently by developing keen observation and using parlay methods. The beauty of the using the parlay is that you do not need the entire group to win.

Let us take a look at one such example.

DC lottery played 718 and 730 on 11/30/2011 followed by 331 and 167 on 12/1/2011. Those Washington DC 3 winning numbers could be won by employing parlay methods based on Figure 10A. You are not going to see those DC 3 winning numbers in Figure 10A.

A good observer would notice the common two digits between 218 in Figure 10A and 718 from DC 3 lottery. You are not expected to be playing the parlay at that point. The candle would be lit at the point DC lottery played 730 that would have two common digits with 830 from this book.

This is the point you would consider playing parlay and the rest of the pick 3 numbers in the group. They came back and played 331 that once again have two common digits with 123 from the book. You would have won that as well as the next winning pick 3 number 167 based on the two common digits (16) from the 681.

You would have won more money by applying the same parlay methods on DC 3 winning number 357 that played on 12/8/2011. You will notice that DC lottery played 187 on 12/7/2011 followed by 030 on 12/8/2011.

The good observer will see the 187 against 218 from this book and 030 against 830. You can see the two common (18 and 30) digits among them and embark on playing parlay from the next winning pick 3 number 751 based on that trend. You will end up winning with the 357 based on the two common digits (57).

You will find those among the groups that cover every lotto market.

There are times you will end up winning not just the parlay but the pick 3 itself thereby giving you a whole lot more.

One of such instances will be by playing the pick 3 numbers and the parlay with the group from Figure 9D against DC 3 results from 1/6/2011 and 1/7/2011. I will pair those results against our manual calculations to show you the opportunities.

The ones with two common digits or more would win money.

DC 3 played 517 on 1/6/2011 and we have 750.

DC 3 played 543 on 1/6/2011 and we have 473.

DC 3 played 139 on 1/7/2011 and we have 113.

DC 3 played 277 on 1/7/2011 and we have 277.

You can see that you would have won the parlay with 13 and then catch the big money maker with 277.

Those numbers are good everywhere including your local lottery market.

You can clearly begin to see that it is possible to win every day.

This is what makes the parlay methods powerful. You would have won two of those games based on the common numbers with the group from 10A.

You will find those all the time by using very good observation.

The explanations and methods in this book are what you are going to use in the other series. It is important that you read this book thoroughly. It is important that you do because the book 2 will give you trends that would rely on the explanations from this book. You should consider reading it more than once.

Chapter 8

Betting With House Money

Let me introduce the term house money and how you can take advantage of it.

The term house in this context is anybody or bodies that represent the lottery in your local market. The term house money represent the surplus money won that could be used to drastically increase your potential winnings.

I will show you an example of winning more with house money with the pick 3 groups in Figure 1A and Figure 2E through New York lottery number results.

The entire groups in this group and the other series are calculated to move in tandem. They groups move in one continuous rotation that enables you to search for and identify the winning numbers based on the trends.

One example of this scenario in play would be from the time New York lottery played 968 on 5/15/2011. You will find that winning number 968 in the second row of Figure 1A. The next winning number in the third row of Figure 1A would be 932 which New York lottery did not play. They played instead the pick 3 winning number 374 on 5/16/2011.

You will now begin to look for 374 in the third row position since 932 that is right below 968 did not play. Your quest would take you to 743 that is in the third row of Figure 2E. This means that the trend that started with 968 from Figure 1A has now shifted to the 743 that is in the third row of Figure 2A.

This equally means that the winning numbers are trending downwards. The next winning pick 3 number based on that trend is 888 that is in the fourth row of Figure 2E right below 743. Please pay close attention to this section because you can make money with it every single day.

The winning number 888 is clearly established as the next number in line to play. You will immediately begin to play the 888. As you may know at this point, I do advocate playing the pick 3 numbers as well as the pick 2 which is the parlay method. The parlay method is meant to cover your cost of betting and more to play with the house money.

The way I would be playing the winning number in this case are as follows,

888

X88

88X

The ones with the X represent the parlay. The parlay pays about $50 for every $1 bet won. I will begin to play those immediately after the 374 that played on 5/16/2011.

New York lottery played 988 on 5/24/2011. I would win $50 for my $1 bet. I am using only $1 for convenience sake. You should always wager on the numbers based on your budget.

You will notice that 988 played in exactly sixteen (16) days from the 374 that played on 5/24/2011.

If you subtract (minus) $16 from the $50 won, you will be left with $34. This $34 is profit or what you could call HOUSE MONEY. The main number 888 is yet to play. I have already won some money by playing the parlay.

I will now begin to play with the $34 house money. I continued betting until New York lottery played 888 on 6/10/2011. You will be amazed to know that the 888 played in exactly 34 days from the last winning number 988 of 5/24/2011.

I started playing with $34 house money and the number played 34 days later. This means that I would win $600 without using my money.

A lot of bettors would be speechless on the possibility of this.

There are numerous things that made it possible.

The winning numbers in the groups move as one unit. They are manually calculated based on trends. You can use each of the groups as mirrors against the other ones. The rotation works continuously thereby capturing the trends at any given moment in every lottery market.

There is no pick 3 lottery that you cannot win if you study the book 1 and 2 very well.

Can you imagine how much more you could win when you collect all the book series?

Thank you!

Lottery Star Book

Made in the USA
Lexington, KY
28 December 2012